My Science Library

Magnet Power

by Buffy Silverman

Science Content Editor:
Kristi Lew

ROURKE CLASSROOM

www.rourkeclassroom.com

Science content editor: Kristi Lew

A former high school teacher with a background in biochemistry and more than 10 years of experience in cytogenetic laboratories, Kristi Lew specializes in taking complex scientific information and making it fun and interesting for scientists and non-scientists alike. She is the author of more than 20 science books for children and teachers.

www.rourkeclassroom.com

Photo credits: Cover © Sideways Design, iadams, Matthew Cole, Cover logo frog © Eric Pohl, test tube © Sergey Lazarev; Page 5 © Matthew Cole; Page 7 © MilanB; Page 9 © Morgan Lane Photography, Kristina Postnikova, Olaru Radian-Alexandru, vision77; Page 11 © Marcel Mooij; Page 12 © April Cat; Page 13 © Blue Door Publishing; Page 15 © Blue Door Publishing; Page 17 © Andrea Danti; Page 19 © Blue Door Publishing, stocknshares; Page 21 © Ilya Andriyanov

Editor: Kelli Hicks

My Science Library series produced for Rourke by Blue Door Publishing, Florida

Library of Congress Cataloging-in-Publication Data

Silverman, Buffy.
 Magnet power / Buffy Silverman.
   p. cm. -- (My science library)
 Includes bibliographical references and index.
 ISBN 978-1-61741-740-5 (Hard cover) (alk. paper)
 ISBN 978-1-61741-942-3 (Soft cover)
 1. Magnets--Juvenile literature. 2. Magnetic pole--Juvenile literature. 3. Magnetism--Juvenile literature. I. Title.
 QC757.5.S55 2012
 538'.4--dc22
                                2011003873

Rourke Publishing
Printed in China, Voion Industry
  Guangdong Province
042011
042011LP

www.rourkeclassroom.com - rourke@rourkepublishing.com
Post Office Box 643328 Vero Beach, Florida 32964

Table of Contents

What Is a Magnet?

Nails dangle from a **magnet**. What holds the nails on the magnet? A **magnetic field** pulls the nails to the magnet.

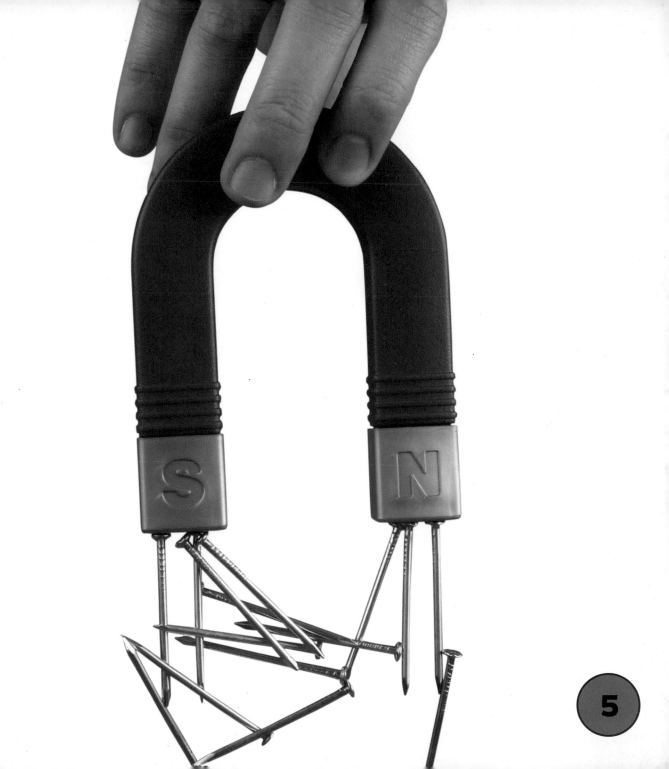

5

A magnet has a **force** around it. You cannot see the force. The force is called a magnetic field.

The black dust on this paper is tiny pieces of iron. The magnetic force pulls the iron pieces to the magnet.

7

All magnets have a magnetic field. The field **attracts** iron, nickel, and certain other metals.

Look at the objects below. Why aren't the block, marbles, and dice stuck to the magnet?

9

Magnets Have Poles

What happens if two magnets are held together? Sometimes they pull together. Sometimes they push apart.

This pyramid was made with yellow magnets and steel balls. The ends of the magnets pull together.

A magnet has two ends, called **poles.** One end is called the north pole (N). The other end is called the south pole (S).

north
pole (N)

south
pole (S)

The north pole of one magnet attracts the south pole of another magnet.

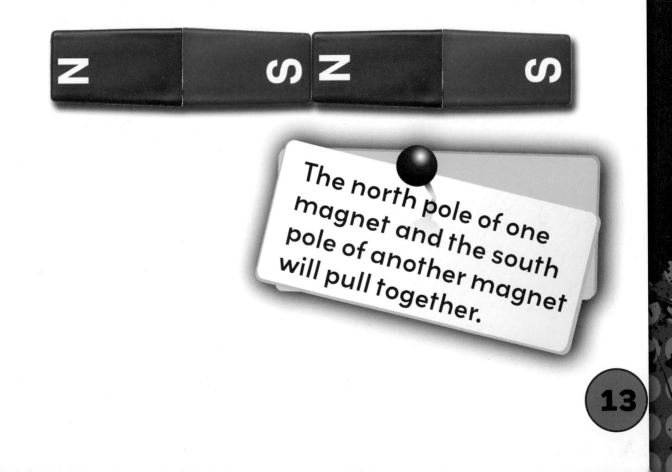

The north pole of one magnet and the south pole of another magnet will pull together.

Poles that are different pull together. Poles that are the same push apart. They **repel** each other.

Can you bring the north ends of two magnets together? No! The magnets push apart.

The Earth Is a Magnet

The Earth acts like a giant magnet. The Earth has two magnetic poles. Its magnetic field stretches out into space.

magnetic
north pole

magnetic field

magnetic field

magnetic
south pole

17

Let a magnet swing on a string. The north pole of the magnet moves. It is pulled toward the Earth's North Pole.

A compass uses the Earth's magnetic field to tell directions. The needle inside is a magnet. If you hold the compass still, the needle will point north.

For hundreds of years people have used compasses to find their way.

Show What You Know

1. Can you name two objects that a magnet picks up?

2. What happens when the south pole of one magnet is near the north pole of another magnet?

3. Why does a compass point north?

Glossary

attracts (uh-TRAKTS): pulls toward

force (FORS): something that pulls or pushes something else

magnet (MAG-nit): an object that attracts iron and has a magnetic field

magnetic field (mag-NEH-tik FEELD): area around a magnet with the power to attract iron

poles (POLES): the two opposite ends of a magnet

repel (rih-PEL): to push away

Index

Websites

www.kids-science-experiments.com/invisibleforce.html

www.sciencebob.com/experiments/electromagnet.php

www.bbc.co.uk/schools/ks2bitesize/science/physical_processes/
    magnet_springs/play.shtml

www.bbc.co.uk/schools/ks2bitesize/science/physical_processes/
    magnet_springs/read1.shtml

http://pbskids.org/zoom/activities/do/funnyfamilyfridgemag.html

About the Author

When Buffy Silverman was a child, one of her prized possessions was a red horseshoe magnet. Now she writes science books for children.